An Operational Process for Workforce Planning

Robert M. Emmerichs

Cheryl Y. Marcum

Albert A. Robbert

Prepared for the Office of the Secretary of Defense

RAND
National Defense Research Institute

The research described in this report was sponsored by the Office of the Secretary of Defense (OSD). The research was conducted in the RAND National Defense Research Institute, a federally funded research and development center supported by the OSD, the Joint Staff, the unified commands, and the defense agencies under Contract DASW01-01-C-0004.

Library of Congress Cataloging-in-Publication Data

Emmerichs, Robert M.
    An operational process for workforce planning / Robert M. Emmerichs,
Cheryl Y. Marcum, Albert A. Robbert.
      p. cm.
    "MR-1684/1."
    Includes bibliographical references.
    ISBN 0-8330-3452-9 (pbk.)
    1. United States. Dept. of Defense—Procurement—Planning. I. Marcum,
Cheryl Y. II. Robbert, Albert A., 1944– III.Title.

UC263.E28 2003
355.6'1'0973—dc22

2003015748

The RAND Corporation is a nonprofit research organization providing objective analysis and effective solutions that address the challenges facing the public and private sectors around the world. RAND's publications do not necessarily reflect the opinions of its research clients and sponsors.

RAND® is a registered trademark.

Cover design by Barbara Angell Caslon

© Copyright 2004 RAND Corporation

Published 2004 by RAND
1700 Main Street, P.O. Box 2138, Santa Monica, CA 90407-2138
1200 South Hayes Street, Arlington, VA 22202-5050
201 North Craig Street, Suite 202, Pittsburgh, PA 15213-1516
RAND URL: http://www.rand.org/
To order RAND documents or to obtain additional information, contact
Distribution Services: Telephone: (310) 451-7002;
Fax: (310) 451-6915; Email: order@rand.org

PREFACE

The Acquisition 2005 Task Force final report, *Shaping the Civilian Acquisition Workforce of the Future* (Office of the Secretary of Defense, 2000), called for the development and implementation of needs-based human resource performance plans for Department of Defense (DoD) civilian acquisition workforces. This need was premised on unusually heavy workforce turnover and an expected transformation in acquisition products and methods during the early part of the 21st century. The Director of Acquisition Education, Training and Career Development within the Office of the Deputy Under Secretary of Defense for Acquisition Reform, in collaboration with the Deputy Assistant Secretary of Defense for Civilian Personnel Policy, asked the RAND Corporation to assist the Office of the Secretary of Defense and several of the defense components in formulating the first iteration of these plans and then evaluating the components' plans.

As part of this project, RAND identified a process, described in this document, that any organization can use to conduct workforce planning. This document is intended to serve as a user's guide for participants conducting workforce planning as they begin to institute the activity in their organization. A companion report, *An Executive Perspective on Workforce Planning*, MR-1684/2-OSD, completes the context for this work and describes the critical role that corporate and line executives play in the workforce planning activity.

This report will be of interest to corporate and business unit executives and to line and functional managers in the DoD acquisition community and to DoD human resource management professionals, as the workforce planning activity continues to mature. In addition, it is oriented and will be more generally of interest to other corporate and business unit executives and to line and functional managers, including human resource professionals—both within and outside the DoD—whose organizations and functions face a similar need for workforce planning.

This research was conducted for the Under Secretary of Defense for Acquisition, Technology, and Logistics and the Under Secretary of Defense for Personnel and Readiness within the Forces and Resources Policy Center of RAND National Defense Research Institute, a federally funded research and development center sponsored by the Office of the Secretary of Defense, the Joint Staff, the unified commands, and the defense agencies.

Comments are welcome and may be addressed to the project leader, Albert A. Robbert at Al_Robbert@rand.org, 703-413-1100, Ext. 5308.

For more information on the Forces and Resources Policy Center, contact the director, Susan Everingham, susan_everingham @rand.org, 310-393-0411, Ext. 7654, at the RAND Corporation, 1700 Main Street, Santa Monica, California 90401.

CONTENTS

FIGURES

SUMMARY

Workforce planning is an organizational activity intended to ensure that investment in human capital results in the timely capability to effectively carry out the organization's strategic intent.[1] This report describes a RAND-developed methodology for conducting workforce planning applicable in any organization. We describe the methodology primarily in terms of its application at a business unit level. We recognize that workforce planning activities can be accomplished at other organizational levels (for example, major divisions within large organizations or even at corporate headquarters). We believe strongly, however, that workforce planning, if not conducted by a business unit itself, nevertheless benefits extensively from the active participation and input of business units.

This report is based on our review of workforce planning in governmental and private-sector organizations and our analysis of the results of the initial application of workforce planning in the DoD acquisition community.[2] In addition to the active involvement of

---

[1]We define *strategic intent* as an expression (sometimes explicit, but often implicit) of what business the organization is in (or wants to be in) and how the organization's leaders plan to carry out that business. Leaders usually express strategic intent in the organization's strategic planning documents. In particular, the business the organization is in (or wants to be in) is often outlined in a vision, mission, and/or purpose statement. How the leaders choose to carry out the business is often captured in goals, guiding principles, and/or strategies. A major task for workforce planners is to identify explicitly those elements of strategic intent that workforce characteristics help accomplish.

[2]Six DoD components completed an initial application of the structured workforce planning process described herein for its acquisition community during the summer of 2001.

business units, we identified three key factors contributing to successful workforce planning:

- enthusiastic executive and line manager participation

- accurate and relevant data

- sophisticated workload models (which help translate expected workloads into requirements for workers) and inventory projection models (which depict how the expected composition of a workforce will change over time).

Different perspectives provide insight into the degree to which these factors influence the effectiveness of the workforce planning activity. Therefore, we structured this report around three points of view: a goal-oriented view—addressing why an organization should conduct workforce planning; a structural view—addressing what questions an organization can answer with workforce planning and the information needed to do so; and a process view—addressing how an organization can effectively focus the contributions of its key participants in conducting workforce planning.

The goal-oriented view sets the stage. It identifies three purposes of workforce planning:

- to obtain a clear representation of the workforce needed to accomplish the organization's strategic intent

- to develop an aligned set of human resource management policies and practices[3]—in other words, a comprehensive plan of action—that will ensure the appropriate workforce will be available when needed

- to establish a convincing rationale—a business case—for acquiring new authority and marshalling resources to implement the human resource management policies and programs needed to accomplish the organization's strategic intent.

---

[3]An aligned set of policies and practices supports the leaders' strategic intent (i.e., the policies and practices are vertically aligned) and are mutually reinforcing (i.e., they are horizontally aligned).

In this context, accomplishing strategic intent is the central goal of workforce planning. The leaders' strategic intent *focuses* workforce planning. Because strategic intent is best defined and articulated by corporate and business unit executives and line managers, the clarity and quality of their input represents a critical factor in successful workforce planning.

The structural view expounds on the purposes of workforce planning. Four central questions capture the major structural themes of workforce planning:

1. What critical workforce characteristics will the organization need in the future to accomplish its strategic intent, and what is the desired distribution of these characteristics?

2. What is the distribution—in today's workforce—of the workforce characteristics needed for the future?

3. If the organization maintains current policies and programs, what distribution of characteristics will the future workforce possess?

4. What changes to human resource management policies and practices, resource decisions, and other actions will eliminate or alleviate gaps (overages or shortages) between the future desired distribution and the projected future inventory?[4]

Figure S.1 shows a blueprint portraying the interrelationship of the questions and the information needed to obtain the answers. It highlights the three key success factors mentioned earlier. Executive and line judgments are critical to questions 1 and 4; data availability is essential to question 2; and a modeling capability is necessary to answer questions 3 and 4.

---

[4]We employ the term *inventory* in this report in the commonplace usage of the DoD to refer to the people in the workforce. For example, *current inventory* represents the number of people currently working in the organization described in terms of such characteristics as length of service, grade, occupation, etc.; *future inventory* represents the number of people expected to be working in the organization at a specified future time described in terms of such characteristics. An *inventory (or workforce) projection model* is an analytic tool for deriving an estimate of the future inventory based on the current inventory.

RAND MR1684/1-S.1

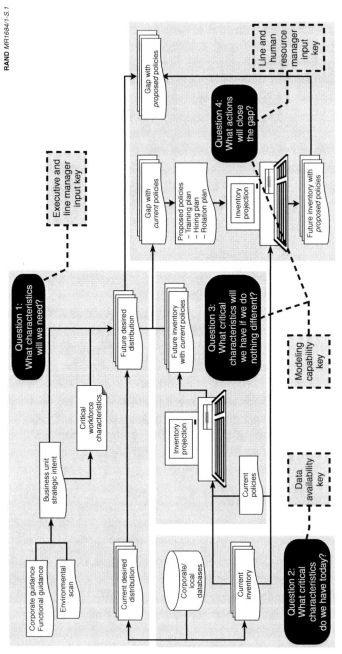

Figure S.1—A Blueprint for Workforce Planning

The process view—the third view presented in this report—operationalizes the blueprint. RAND proposes a four-step process any organization can use to focus the contributions of its key participants in conducting workforce planning. We designed this process for application at the business unit level. Figure S.2 outlines the four steps. We envision the process as a structured dialogue among the business unit's senior leaders: its executives, line managers, community managers,[5] and human resource managers. Executive and line manager participation is critical to steps 1 and 2; line, community, and human resource manager participation is critical to step 4. The process relies on comprehensive data and sophisticated models to ensure that participants can effectively accomplish steps 2 and 3.

RAND *MR1684/1-S.2*

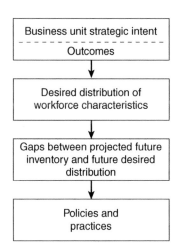

Figure S.2—A Four-Step Workforce Planning Process

---

[5]Many organizations assign career development and other human resource–related responsibilities for individuals in specific occupational or professional groups to senior executives in the occupation or professional group. In addition, senior executives often oversee these types of responsibilities for individuals working in major functional areas (such as acquisition or finance). These *community managers* (or *functional community managers*) are expected to ensure that the workforce possesses the capabilities needed by business units.

RECOMMENDATIONS

Importantly, an organization's senior leaders control the three key factors that lead to successful workforce planning.

Active Executive and Line Participation. Business unit executives and line managers are uniquely positioned to assess how their business will be carried out and to identify the human capital capabilities needed to do so effectively. Leaders at levels above the business unit play different but important roles—in translating higher-level direction into clear guidance for line organizations, integrating the results of workforce planning across business units, and supporting the results of workforce planning at the lower levels. We recommend that senior leaders (above the business unit level) delineate explicitly the roles and responsibilities of executives involved throughout the organization in the workforce planning activity, and in particular, encourage and reward business unit executives and line managers for active participation in the workforce planning activity.

Accurate and Relevant Data. Data on workforce characteristics are the common language of workforce planning. Although many facets of workforce planning are best carried out by individual business units, the kind of data needed is similar throughout the organization (across business units). We recommend that the organization's corporate headquarters lead the development of the functional specifications for a human resource information system to support workforce planning.

Sophisticated Workload and Inventory Projection Models. Insight into how the composition of the workforce may change over time informs human capital decisions. As with data, the kind of models needed to make such estimates may be similar throughout the organization. We recommend that the organization's corporate headquarters evaluate the availability, costs, and benefits of comprehensive, integrated workload and inventory projection models for all categories of employees (including contractors, where contractor personnel are integrated with civil service employees).

ACKNOWLEDGMENTS

The research underlying this report had its genesis in the Office of the Deputy Assistant Secretary of Defense for Civilian Personnel Policy and the Office of Acquisition Education, Training and Career Development and Director of the Acquisition Workforce 2005 Task Force, in the Office of the Assistant Secretary of Defense for Acquisition Reform. They recognized the need for better workforce planning capabilities, particularly with respect to Defense acquisition workforces, and committed resources to providing those capabilities, including sponsorship of our research and assistance. We received valuable advice and assistance from many individuals within these offices,

Part of our research took us into close contact with two acquisition business units—the Space and Naval Warfare Systems Command (SPAWAR) and the Naval Facilities Engineering Command (NAVFAC). The leadership of Rear Admiral Kenneth Slaght (SPAWAR) and Rear Admiral Michael Loose (NAVFAC) was one of the most valuable contributions to our research. A number of individuals in those commands helped us make our consultations productive, including Margaret Malowney, Director of Human Resources at SPAWAR; Margaret Craig, Executive and DAWIA Training Coordinator at SPAWAR; Amy Younts, Director of Community Management at NAVFAC; Sara Buescher, Director of Civilian Personnel Program, NAVFAC; Joy Bird, Associate Director of Community Management, NAVFAC; and Hal Kohn, Senior Systems Analyst for Community Management, NAVFAC.

This research builds on the conceptual foundation of strategic human resource management propounded by the presidentially chartered Eighth Quadrennial Review of Military Compensation in 1997, which was further refined and applied by the Naval Personnel Task Force, convened in 1999 by the Secretary of the Navy and the Assistant Secretary of the Navy for Manpower and Reserve Affairs.

RAND colleague Harry Thie and Steve Kelman, Professor of Public Management at the Kennedy School of Government, Harvard University, provided thoughtful reviews. Miriam Polon edited the manuscript. Any remaining errors are, of course, our own.

ACRONYMS

DAWIA   Defense Acquisition Workforce Improvement Act

DoD   Department of Defense

DON   Department of the Navy

GAO   U.S. General Accounting Office

NAVFAC   Naval Facilities Engineering Command

OSD   Office of the Secretary of Defense

SPAWAR   Space and Naval Warfare Systems Command

YOS   year of service

Chapter One

INTRODUCTION

Workforce planning is neither a new activity nor an ephemeral management craze. Organizations in the private sector have engaged in workforce planning for decades.[1] It is, however, a relatively recent activity throughout most of the federal government, with some notable exceptions such as planning for military forces by the military services in the Department of Defense (DoD). Although a large number of impending retirements stimulated interest in civilian workforce planning in the DoD and many other federal agencies,[2] some agencies have taken the next step and incorporated workforce planning into their overall planning process.[3]

Unfortunately, widespread discussion of workforce planning within the federal government has been mostly about why an organization should conduct workforce planning. Little discussion specific enough to help organizations develop what questions an organization can answer with workforce planning and the information needed to do so or how an organization can effectively focus the contributions of its key participants in conducting workforce plan-

---

[1]The Corporate Leadership Council profiled the workforce planning efforts of nine private sector corporations in two Fact Briefs (1997 and 1998). The findings outlined in the Fact Briefs are consistent with the methodology described in this report and particularly emphasize the importance of incorporating business strategy in the workforce planning process.

[2]*Federal Employee Retirements: Expected Increase Over the Next 5 Years Illustrates Need for Workforce Planning*, GAO-01-509, April 2001.

[3]The Office of Personnel Management lists workforce planning activities (and contacts) within the federal government at http://www.opm.gov/workforceplanning/WhosDoingWhat.asp, accessed June 18, 2003.

ning has occurred. Even in those organizations that have developed a workforce planning process, the process is often uniquely embedded in the overall planning process and is not easily adaptable to other organizations.

More serious in our view, where workforce planning is practiced in the federal government, executive input and organizational strategies seldom directly influence the workforce planning activity. In a dynamic environment, we believe the workforce planning activity is more effective—and more likely to be worth the resources needed to carry it out—if linked overtly to the organization's strategic intent.[4] We also believe this link is clearest at the business-unit or business line level, especially in a large, complex federal agency like the DoD. However, we found no organization with an easily transferable method for unambiguously integrating the elements of its strategic intent into its workforce planning process. This may seriously limit the payoff of this activity. In our view, an explicit consideration of the nature of the business the organization wants to be in and how it wants to carry out that business is what makes human capital planning—and workforce planning—strategic.[5]

To overcome existing shortcomings and to emphasize the centrality of strategic intent in particular, RAND developed a comprehensive workforce planning methodology. The methodology incorporates much of what is frequently described in discussions of workforce planning in the literature and applied in public- and private-sector organizations. Importantly, however, we structured this methodology to explicitly link decisions about human resource management policies and practices to an organization's strategic intent at the business unit level. Although the general notion of linking human re-

---

[4]We define *strategic intent* as an expression (sometimes explicit, but often implicit) of what business the organization is in (or wants to be in) and how the organization's leaders plan to carry out that business. Leaders usually express strategic intent in the organization's strategic planning documents. In particular, the business the organization is in (or wants to be in) is often outlined in a vision, mission, and/or purpose statement. How the leaders choose to carry out the business is often captured in goals, guiding principles, and/or strategies. A major task for workforce planners is to identify explicitly those elements of strategic intent that workforce characteristics help accomplish.

[5]See the companion report, *An Executive Perspective on Workforce Planning*, MR-1684/2-OSD, for that discussion.

source management policies and practices to an organization's strategic intent is not original, our methodology is distinctive in that it offers an organization a systematic means of explicitly identifying and illustrating that linkage. We believe this is a major contribution to workforce planning—and more generally to human capital strategic planning.

This report describes this workforce planning methodology from three points of view: a goal-oriented view, a structural view, and an organizational-process view. Although the three views present much the same information, the information is organized in slightly different ways. We intend each point of view to emphasize different important aspects or dimensions of workforce planning.

- The goal-oriented view addresses the *why* of workforce planning: It describes three purposes of workforce planning that generally characterize any such planning effort. An organization's leaders may find this point of view useful in assessing whether it wants to engage in workforce planning.

- The structural view addresses the *what* of workforce planning: It identifies four questions that capture the major themes of workforce planning, and describes the information needed to answer those questions. The answers are related; in fact, they build on each other. The structural view integrates the four questions and, thereby, provides a blueprint for conducting the workforce planning activity. An organization's leaders may find this point of view useful in developing a shared picture of the mechanics of workforce planning (the major elements and how they fit together).

- Finally, the organizational-process view—the centerpiece of the methodology—addresses the *how* of workforce planning. It describes a four-step organizational process to engage key participants in operationalizing the blueprint. This four-step workforce planning process encompasses three key ingredients: an explicit identification and consideration of an organization's strategic intent, an incremental approach to estimating the future desired distribution of workforce characteristics, and a model for projecting the distribution of workforce characteristics the workforce is expected to possess in the future. An organization's leaders may find this point of view useful in identifying and mobilizing the resources needed to conduct the major activities of workforce planning.

Importantly, this organization-centric methodology applies to any organization at the business unit level. We believe participants in the workforce planning activity at the business unit level are in the best position to understand human capital[6] needs and to identify the human resource management policies and practices—the leadership tools—required to ensure that those needs are met.[7]

ORGANIZATION OF THE REPORT

Chapters Two through Four discuss, respectively, the goal-oriented, structural, and organizational-process views of workforce planning described above. Chapter Five makes recommendations for bolstering the key success factors underlying the workforce planning methodology. Appendix A describes a RAND-developed workforce projection model, and Appendix B contains a sample agenda for conducting the workforce planning process.

---

[6]Throughout this report, we use the term "human capital" in the same way the U.S. General Accounting Office (GAO) uses it: "In contrast with traditional terms such as personnel and human resource management, it focuses on two principles that are critical in a performance management environment. First, people are assets whose value can be enhanced through investment. As the value of people increases, so does the performance capacity of the organization, and therefore its value to clients and other stakeholders. Second, an organization's human capital approaches must be aligned to support the mission, vision for the future, core values, goals, and strategies by which the organization has defined its direction and its expectations for itself and its people." (GAO, 2000.)

[7]Consequently, if an organization chooses to centralize workforce planning at a level above the business unit, we strongly believe business unit representation, participation, and input are critical to success.

Chapter Two

A GOAL-ORIENTED VIEW: THE PURPOSE OF STRATEGIC WORKFORCE PLANNING

The definition of workforce planning affords insight into its purpose. For example, the National Academy of Public Administration (2000, p. 1) defines strategic workforce planning as "a systematic process for identifying the human capital required to meet organizational goals and for developing the strategies to meet these requirements." Ripley (2000, p. 1) suggests that workforce planning is "a systematic assessment of workforce content and composition issues and [determination of] what actions must be taken to respond to future needs."

Building on these and similar definitions, we concluded that organizations employ workforce planning to accomplish at least three purposes:

- to obtain a clear representation of the workforce needed to accomplish the organization's strategic intent

- to develop an aligned set of human resource management policies and practices[1]—in other words, a *comprehensive* plan of action—that will ensure that the appropriate workforce will be available when needed[2]

---

[1]An aligned set of policies and practices supports the leaders' strategic intent (i.e., the policies and practices are vertically aligned) and are mutually reinforcing (i.e., they are horizontally aligned).

[2]Other reports and practitioners use terms such as "actions," "strategies," and "initiatives"; this report focuses explicitly on "an aligned set of human resource man-

- to establish a convincing rationale—a business case—for acquiring new authority and marshalling resources to implement the human resource management policies and programs needed to accomplish the organization's strategic intent.

These purposes respond to the *why* of workforce planning. They are expressed in terms of the major results an organization can expect when it engages in this activity. Importantly, a common theme cuts across the three purposes: The focal point of workforce planning is the organization's ability to accomplish its strategic intent.

---

agement policies and practices" as the product of strategic human capital decisions. We view policies and practices as primary leadership tools for shaping the workforce.

Chapter Three

A STRUCTURAL VIEW: THEMES AND A BLUEPRINT FOR WORKFORCE PLANNING

This chapter presents the structural view of workforce planning. It describes a comprehensive conceptual foundation upon which an organization can build when engaging in the workforce planning process at the business unit or business line level. To establish this foundation, we first pose four thematic questions any operational workforce planning process must answer. We then use these questions to construct a workforce planning blueprint that describes the essential components of workforce planning. To convert the workforce planning blueprint into reality, however, an organization needs a process for engaging its key participants. Chapter Four describes that organizational process.

FOUR THEMATIC QUESTIONS

Planning sets out to address specific types of issues or to answer particular kinds of questions. The National Academy of Public Administration (2000) highlights four issues of the type workforce planning can address:

- The composition and content of the workforce that will be required to strategically position the organization to deal with possible future situations and business objectives

- The gaps that exist between the future "model" organization and the existing organization, including any special skills required for possible future situations

7

- The recruiting and training plans for permanent and contingent staff that must be implemented to deal with those gaps

- The determination of what functions or processes, if any, should be outsourced, and how.

The Office of Management and Budget's direction (OMB, 2001, p. 2) regarding workforce planning is representative of the kinds of questions organizations seek to address through workforce planning:

- What skills are currently vital to the accomplishment of the agency's goals and objectives?

- What changes are expected in the work of the agency (e.g., as a result of changes in mission or goals, technology, new or terminated programs or functions, and shifts to contracting out)? How will these changes affect the agency's human resources? What skills will no longer be required, and what new skills will the agency need in the next five years?

- What recruitment, training, and retention strategies are being implemented to help ensure that the agency has, and will continue to have, a high-quality, diverse workforce?

- How is the agency addressing expected skill imbalances resulting from attrition, including retirements over the next five years?

- What challenges impede the agency's ability to recruit and retain a high-quality, diverse workforce?

- Where has the agency successfully delegated authority or restructured to reduce the number of layers that a programmatic action passes through before it reaches an authoritative decision point (e.g., procuring new computers, allocating operating budgets, completely satisfying a customer's complaint, processing a benefit claim, clearing controlled correspondence)? Where can the agency improve its processes to reduce the number of such layers?

- What barriers (statutory, administrative, physical, or cultural) to achieving workforce restructuring has the agency identified?

Although these (and other) sources are informative, they often fail to capture the full range of workforce planning considerations. For example, the OMB guidance and many other discussions of workforce planning focus on skills as the only workforce characteristic of interest. Although skills are often likely to be of interest—perhaps even of primary interest—they are only one of the potential workforce characteristics critical to organizational success. More fundamentally, other discussions leave out what we consider to be important structural components in conducting workforce planning. To overcome some of the deficiencies in existing sources, we developed four thematic questions an organization conducting workforce planning must answer:

1. What critical workforce characteristics will the organization need in the future to accomplish its strategic intent, and what is the desired distribution of these characteristics?

2. What is the distribution—in today's workforce—of the workforce characteristics needed for the future?

3. If the organization maintains current policies and programs, what distribution of characteristics will the future workforce possess?

4. What changes to human resource management policies and practices, resource decisions, and other actions will eliminate or alleviate gaps (overages or shortages) between the future desired distribution and the projected future inventory?[1]

Unfortunately, many practitioners omit the third question. As the next subsection demonstrates, this critical failing can prevent a comprehensive assessment of the fourth question.

Central Concepts

Two central concepts are embedded in the four thematic questions: workforce characteristics and the distribution of workforce characteristics.

---

[1]We employ the term *inventory* in this report in the commonplace DoD usage to refer to the people in the workforce. For example, *current inventory* represents the number of people currently working in the organization described in terms of such characteristics as length of service, grade, occupation, etc.; *future inventory* represents the number of people expected to be working in the organization at a specified future time described in terms of such characteristics.

A *workforce characteristic* is a concrete and measurable aspect of a group of workers that is critical for organizational success and can be influenced by human resource management policy decisions. Examples of workforce characteristics include occupation/job series, experience,[2] competencies or skills (for example, leadership or multifunctionality), and education (for example, degree and discipline).

The *distribution of workforce characteristics* is the frequency of occurrence of a workforce characteristic within an organization. Distribution can be expressed as the number or percentage of individuals (inventory) or positions (requirements) distributed across the categories defining the workforce characteristic. For example, an organization interested in clusters of job series as a workforce characteristic might describe its current inventory as consisting of 40 percent of the workforce in the scientist and engineering cluster, 21 percent in the information technology cluster, 9 percent in the logistics cluster, and the remaining 30 percent in an "all others" cluster. The distribution of workforce characteristics can also be expressed in two or more dimensions. Continuing the example, the science and engineering cluster might have 15 percent of its population with less than 5 years of experience, 37 percent with 5 to 20 years of experience and the remaining 48 percent with over 20 years of experience. Other organizational clusters may have different distributions of experience. As part of the workforce planning process, the organization assesses the desired distribution along the same dimensions as it assesses its inventory. For example, the desired distribution of experience in the categories used above for scientists and engineers might be 20 percent, 50 percent, and 30 percent.

Figure 3.1 portrays a three-dimensional distribution of workforce characteristics, in this case occupation (aggregated by individual job series), educational level (aggregated by degree) and years of service (aggregated into three unequal years-of-service groupings).

---

[2]Given the kind of data currently available in many federal workforce inventory databases, years of service or grade (concrete and measurable, but difficult to argue as critical for organizational success) might serve as a proxy for occupational experience (which is more likely to be identified as critical, but is not systematically measured).

RAND *MR1684/1-3.1*

| Occupational series | Education | | | Years of service | | | | | |
| | | | | 1–3 | | 4–10 | | >10 | |
	Level	Number	%	Number	%	Number	%	Number	%
Program management	<Bachelor	64	31%	3	5%	9	14%	52	81%
	Bachelor	96	46%	1	1%	7	7%	88	92%
	Master/ Professional certification	48	23%		0%	4	8%	44	92%
	Total	208		4	2%	20	10%	184	88%
Mechanical engineer	<Bachelor	22	3%	11	50%	1	5%	10	45%
	Bachelor	552	68%	38	7%	103	19%	411	74%
	Master/ Professional certification	214	26%	17	8%	43	20%	154	72%
	Doctor	26	3%	5	19%	9	35%	12	46%
	Total	814		71	9%	156	19%	587	72%

Legend: Workforce characteristics   Levels of aggregation

Figure 3.1—An Example of a Three-Dimensional Distribution of Workforce Characteristics

With these definitions in mind, we consider each of the four thematic questions posed above.

A BLUEPRINT FOR INTEGRATING AND ANSWERING THE FOUR QUESTIONS

The four questions provide a broad outline of the structure an organization needs to carry out workforce planning. However, many descriptions of workforce planning end here, leaving indeterminate the vital information a practitioner needs to answer the questions. As noted earlier, the four questions are related; all must be answered to produce a meaningful result.

Figure 3.2 portrays a blueprint for integrating and eventually answering the four central questions. It presents a comprehensive picture of how the four questions fit together. The arrows in the figure high-

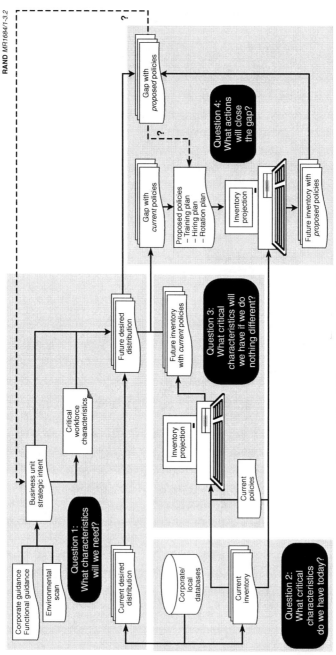

Figure 3.2—A Blueprint for Workforce Planning

light the interrelationships among the four questions primarily in terms of the flow of information.[3]

The remainder of this chapter describes this blueprint and the interrelationships among the four questions.

Future Desired Distribution

1. *What critical workforce characteristics will the organization need in the future to accomplish its strategic intent, and what is the desired distribution of these characteristics?*

Although a challenging task, identifying critical workforce characteristics and describing the future desired distribution of those characteristics is essential. Armed with this information, the organization can narrow its focus and resources to the portion of the workforce that most directly contributes to organizational success, and can clearly articulate a specific target toward which to shape the workforce. Strategic intent is a primary factor in answering this question. In addition, explicitly describing the link between an organization's strategic intent and its critical workforce characteristics strengthens the argument (the business case) for the requisite changes to human resource management policies and practices and for the program resources to implement them.[4]

---

[3]For brevity, throughout this report (and in particular, in Figure 3.2 and those following), consistent with the description of workforce characteristics in the previous section, we use the labels "current inventory," "future inventory," "current desired distribution," and "future desired distribution" to refer to various distributions of critical workforce characteristics. For example, *current inventory* is shorthand for "the distribution of critical workforce characteristics in the current inventory;" *future desired distribution* is shorthand for "desired distribution of critical workforce characteristics in the workforce at a specified future date."

[4]The DoD components engaged in the initial application of workforce planning found it difficult to identify, express, and use strategic intent, particularly in answering question 1. Although they took a significant first step toward implementing workforce planning, they failed to achieve the purposes of workforce planning outlined above—precisely because strategic intent is central to each purpose. However, two "business units" within the Department of the Navy (the Space and Naval Warfare Systems Command [SPAWAR] and the Naval Facilities Engineering Command [NAVFAC]) were successful in this task. In addition, these business units converted their expression of strategic intent into an assessment of the workforce characteristics needed to achieve their strategic intent. In effect, they envisioned how they were going

Strategic intent manifests itself in various forms at different levels of an organization. Sometimes it is implicit, but often it is contained in statements such as an organization's purpose and mission or in a description of what the organization's leaders want the organization to become in the future (their vision) and their general strategy for getting there.

At the business unit level, strategic intent is usually the most concrete; it often includes a specific focus on *how* the business unit leaders intend to carry out their purpose, mission, and vision. In addition, a business unit's strategic plan usually describes its specific goals and the actions its leaders intend to take to realize their strategic intent.

At higher organizational levels of large, complex organizations, strategic intent is generally broader in scope. It focuses not only on how leaders at that level intend to carry out their purpose, mission, and vision but also on what other parts of the organization will do to help the leaders accomplish their mission. Depending on the perspective, the higher level can communicate its strategic intent to lower levels of the organization by issuing *corporate guidance* that affects multiple functions or more narrowly drawn *functional guidance*. Corporate and functional guidance may take a variety of forms, for example: redefinition, realignment, or reprioritization of subordinate missions (resulting, perhaps, from a shift in corporate or functional strategy); desired concrete results expected from subordinate units (performance targets); specific policies to be implemented (because the policies support, say, economies of scale); or specific planning assumptions for subordinate units to use (to ensure a common basis for inter-unit comparisons).

In the context of workforce planning, strategic intent—as manifested, say, in the business unit's strategic plan, informed and supplemented by corporate and functional guidance and a scan of the environment[5]—determines the number and characteristics of people

---

to use that workforce to accomplish their goals. Based on our detailed interaction with these two business units, we concluded that an organization *can*, with firm resolve, use workforce planning to help it achieve its strategic intent.

[5]We define the *environment* as external factors that impact the organization but over which the organization has little or no control.

the business unit needs to accomplish that strategic intent. Although the business unit's strategic intent may be more detailed, corporate or functional guidance could have a greater influence on the workforce characteristics during periods of dramatic shifts in corporate strategy. Consequently, in the blueprint we show both business unit strategic intent and corporate and functional guidance as the principal bases for identifying critical workforce characteristics and assessing the future desired distribution of those characteristics.[6]

An organization can better assess future desired distribution if it has a solid foundation from which to start. Identifying and describing the current desired distribution of critical workforce characteristics can provide such a base. In addition, this process allows an assessment of the effect of a change in the business unit's strategic intent or a change in corporate or functional direction. Figure 3.3 illustrates how current desired distribution, perhaps derived from information in the organization's databases, might be combined with a specific expression of strategic intent to derive future desired distribution.

The Current Inventory

2. *What is the distribution—in today's workforce—of the workforce characteristics needed for the future?*

The answer provides the foundation upon which to build the workforce of the future. As implied in Figure 3.4, some data to answer this question may reside in existing databases. For example, the distribution of occupations is readily available. However, although information may be available for some of the critical characteristics, current databases may not contain information related to all workforce characteristics workforce planners deem critical to accomplishing the organization's strategic intent. For example, few organizations main-

---

[6]An organization normally identifies potential changes in the environment (for example, advances in technology, a shifting political-military situation, or industrial restructuring) in its overall strategic planning process and reflects the effect in corporate and functional guidance and business unit strategic intent. Such changes could influence the future desired distribution of critical workforce characteristics. Such information is relevant to workforce planning when addressing the first thematic question.

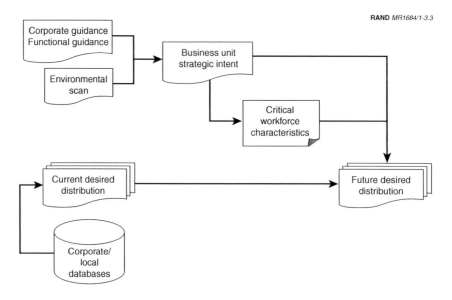

Figure 3.3—Deriving Future Desired Distribution

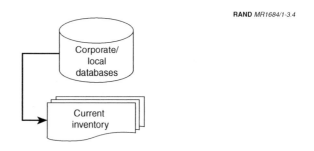

Figure 3.4—Deriving the Current Inventory

tain data on each employee's competencies or skills, characteristics that may become critical in the future for particular organizations. Question 2 highlights the data needed to make informed human capital decisions and, if the data are not currently available, can stimulate efforts to gather it.

The Projected Future Inventory

3. *If the organization maintains current policies and programs, what distribution of characteristics will the future workforce possess?*

The distribution of workforce characteristics in the inventory changes over time because of managerial actions and employee decisions. To account for these actions and decisions, workforce planners use computer models to project the current inventory into the future to estimate how the distribution of workforce characteristics will change. For example, the current workforce will age: Some of those reaching retirement eligibility will leave the workforce (taking important characteristics with them) and new hires will enter (providing the opportunity to select based on desired characteristics or to develop desired characteristics through training and assignments).

Some of the relationships the model simulates are simple and straightforward (for example, each year individuals who remain gain an additional year of service); others are complex, to better reflect reality (for example, loss rates may increase as time-to-promotion to the next grade lengthens). In addition, explicitly incorporating the estimated effects of recently implemented or planned human resource management policies and practices in the model provides a better estimate to inform human capital decisions.

As illustrated in Figure 3.5, an inventory projection model provides a structured, explicit means of answering question 3.

Ideally, quantitative data such as historical trends (for example, continuation and hire rates) are the basis for these projections; however, in many cases, workforce planners may have only a qualitative estimate based on informed judgment upon which to make the projections. In either case, it is important to make the assumptions underlying the projections explicit, to test the validity of the assumptions over time and to assess the robustness of the results.

RAND *MR1684/1-3.5*

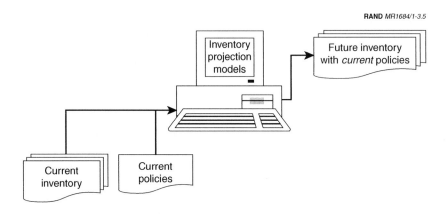

Figure 3.5—Deriving the Expected Future Inventory

Human Resource Management Policies and Practices to Eliminate or Alleviate Gaps

4. *What changes to human resource management policies and practices, resource decisions, and other actions will eliminate or alleviate gaps (overages or shortages) between the future desired distribution and the projected future inventory?*

As depicted in Figure 3.6, comparing the distribution of characteristics in the projected future inventory to the future desired distribution identifies potential shortages and overages—gaps between the future desired distribution and the projected future inventory.

Figure 3.7 shows how identified gaps direct the search for remedial policies and practices. For example, an organization might seek to eliminate shortages in a particular characteristic (such as an occupation) through such human resource management policies and practices as targeted hiring, increased training or development, and/or targeted retention bonuses or by restructuring the organization or the work. Or it might alleviate overages through such policies and

RAND MR1684/1-3.6

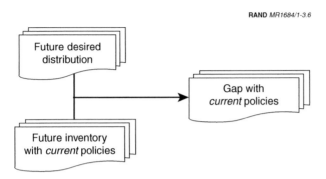

Figure 3.6—Identifying Potential Gaps

practices as early retirement programs or retraining employees to acquire different characteristics.[7]

Human resource management policies and practices are the tools managers use to shape the workforce. Policies and practices designed to ensure the workforce required to accomplish an organization's strategic intent is available when needed are *vertically aligned* with the organization's mission or strategy. For example, line managers may offer a lump-sum payment to retirement-eligible individuals with skills no longer needed in the workforce to give them an incentive to separate; or they may use a tuition reimbursement program as an incentive for individuals with critical skills and several years of experience to join the organization. Policies and practices designed to work in harmony with each other are *horizontally aligned*. For example, a training policy that teaches employees how to collaborate and contribute in a team environment complements a performance measurement policy that measures team performance, which complements a compensation policy that rewards team per-

---

[7]Of course, the final set of policies and practices the organization selects for implementation will likely depend on other considerations besides the ability to eliminate the gap. For example, the organization may choose to prioritize policies and practices on the basis of cost, risk, organizational culture, employee reaction, etc., as well as effectiveness. We discuss these considerations in the next chapter.

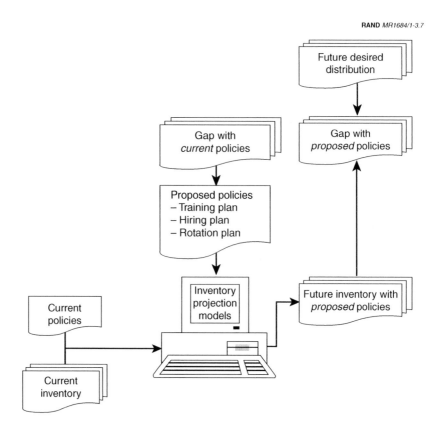

RAND *MR1684/1-3.7*

Figure 3.7—Remedial Policies and Practices to Address Gaps

formance. Ideally, human resource management policies and practices are aligned vertically and horizontally.

In total, the proposed set of policies and practices—aligned and sequenced—comprises the workforce planning input to an organization's human capital strategic plan.

To set the stage for evaluation of the success of the human capital strategic plan, the workforce planners should estimate the expected result of the combined effects of the different policies and practices. Ideally, they would do this by making their underlying assumptions about these effects explicit (quantitatively or qualitatively) and by using these new or additional assumptions to project the current inventory into the future, assuming implementation of the proposed policies and practices.

Comparing the revised projection of the distribution of the workforce characteristics in the future inventory with the future desired distribution permits the workforce planners to identify any remaining gap. The lack of a substantial gap suggests that the human capital strategic plan is comprehensive and effective. Alternatively, a persistent gap suggests the need for additional or different policies and practices or the possibility that the organization cannot accomplish its strategic intent as presently stated. In the latter case, the organization must call for a reevaluation of corporate or functional guidance or the way the business unit has chosen to carry out its mission. The potential requirement for the workforce planners to iterate at least part of the workforce planning process—by revisiting the policies and practices or by reviewing the fundamental way in which the leaders choose to carry out their business—is reflected by the dashed lines in the overall blueprint illustrated in Figure 3.2.

CRITICAL CONSIDERATIONS

Our blueprint highlights the critical role of three considerations often overlooked by organizations conducting workforce planning: executive and line judgments, data, and modeling. We discuss each in turn.

Executive and Line Judgments

Workforce planning is, at its best, a participative activity. Broad participation enhances the quality of the results of workforce planning and generates a coalition of support for implementing those results. It is particularly important (and an area in which many efforts fall short) that the activity capitalize on executive and line participation.

Executive and line input is critical to a comprehensive answer to question 1 (What will we need in the future?). Executives must articulate strategic intent clearly and in terms that can be translated into human capital considerations. Line managers must identify the workforce characteristics and assess the distribution of those characteristics that are needed to accomplish their strategic intent. This is an unfamiliar, and therefore difficult, task for both. It is, in our experience, not primarily an introspective task. Rather, it benefits from a dialogue among the executives and the line managers—in fact, in many instances it may only be accomplished through dialogue. An important side benefit is a better shared understanding of where these senior leaders of the organization want to go, how they want to get there, and what the human capital implications of that intent are.

A comprehensive answer to question 4 (How do we preclude gaps and overages?) relies critically on input from line managers, together with input from community managers (when such a role exists in the organization).[8] Line managers employ policies and practices (for example, selection, assignment, rewards, performance measurement) to acquire, retain, and separate individuals based on targeted workforce characteristics. Community managers, as well as line managers, employ policies and practices (for example, training and development) to acquire and sustain human capital. Line and community managers provide "customer" input for developing policies and practices. This participation also benefits the eventual implementation of policies and practices.

Although an organization will find no aspect of workforce planning more difficult to conduct than the executives and line managers determining the workforce characteristics that are critical to accomplishing their strategic intent and the future desired distribution of those characteristics, it will also find no aspect that adds more value to the result.

Data

The blueprint uses data as the lifeblood of workforce planning. Two forms of data are critical: stocks and flows.

---

[8]Many organizations assign career development and other human resource–related responsibilities for individuals in specific occupational or professional groups to senior executives in the occupation or professional group. In addition, senior executives often oversee these types of responsibilities for individuals working in major functional areas (such as acquisition or finance). These *community managers* (or *functional community managers*) are expected to ensure that the workforce possesses the capabilities needed by business units.

Workforce characteristics and their distributions represent stocks. Workforce planners collect most of their data as stocks. Some of these data are centralized in agency-wide databases.
Data as stocks are required to portray the current inventory, that is, to answer question 2 (What do we have today?). Problems ensue when data on a critical workforce characteristic are not available (as may be the case for acquisition experience) or when available data inaccurately reflect the manifestation of the characteristic actually in the workforce (as may be the case for educational level achieved).

Data as stocks can assist in assessing the future desired distribution of a critical workforce characteristic as well, contributing to the answer to question 1 (What will we need in the future?). This assistance can be direct, as in the case of existing data on the *current* desired distribution—providing executives and line managers a baseline from which to make judgments about the future desired distribution. It can also be indirect. If no data on the current desired distribution are available, the data provide a base (current inventory) for incrementally estimating current desired distribution and then future desired distribution.[9]

Continuation rates, hiring rates, and transition rates between states (for example, from the state of having a bachelor's degree to having a master's degree) represent flows. Data as flows are generally less available than data as stocks, and they are more difficult to collect.

Data as flows are required to build and calibrate the models required to answer question 3 (What will we have if we maintain current policies?) and to support assessment of proposed policies and practices in answer to question 4 (How do we preclude gaps and overages?). Assessing the effects of proposed policies and practices may initially rely on judgmental estimates of changes to the flows (using panels, focus groups, Delphi techniques, etc.). The value of such estimates lies in making them explicit—and thereby subject to inspection and eventual validation or modification.

So workforce planning relies heavily on data. However, the lack of a complete set of data (stocks and flows) should not prevent the organization from conducting workforce planning. Planners will always

---

[9]We describe this incremental process in Chapter Four.

have enough data to *begin* the workforce planning activity. However, they will require more data to support the activity as it matures. The workforce planning activity itself provides a structured approach for identifying areas where the planners need to devote effort and resources to acquiring or refining the data.

Figure 3.8 portrays a general approach for identifying available data to begin a workforce planning cycle, and focusing the data-gathering activity to refine inaccurate data or to develop unavailable data for use in successive workforce planning cycles. It also highlights the requirement to collect data on the assumed effects of the decisions to implement revised policies and practices, so the workforce planners can evaluate and revise those decisions as necessary to better achieve the organization's strategic intent. For example, the planners assume that implementing a new on-the-spot hiring policy will enable the organization to hire an additional 100 engineers within four months. If, at the end of four months, the organization had hired only 50 additional engineers, it would be necessary to revise their initial assumption—and to investigate additional policies and practices to make up the hiring shortfall.

RAND *MR1684/1-3.8*

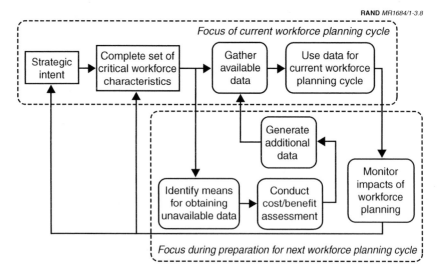

Figure 3.8—General Approach for Focusing the Data-Gathering Activity

Modeling

The blueprint highlights the critical role of modeling in supporting workforce planning. Modeling is central to answering question 3 (What will we have if we maintain current policies?), especially because we have formulated this question differently than other practitioners have. Others suggest that workforce planning should focus on the difference between the distribution of the characteristic in the current inventory and the future desired distribution, but their approach does not require workforce planners to estimate how the composition of the current workforce will change over time. We argue that workforce planners have two important reasons to evaluate the change in the current workforce composition over time.

First, managers generally will have applied policies and practices in the past to respond to existing issues and problems. Managers may have implemented some policies and practices in the remote past, others just recently. The effects of the former are potentially well known, whereas the latter may not have reached full effectiveness. This mix of policies and practices will influence the composition of the workforce over time, possibly in a complicated way. Decisions regarding further changes will be ill-founded if workforce planning participants do not account for the unrealized effects of decisions already made.

Second, current policies and practices may have been implemented to meet current requirements or simply to replace the current force. Even if this intent is realized, further changes may be needed to reflect the difference between the distribution of workforce characteristics in the expected future inventory (assuming current policies and practices will continue) and what the future desired distribution will require.

Assuredly, in relatively simple workforce planning activities planners can make these estimates without using a model. As an activity grows in complexity, however, such estimates generally will not be possible. As with data, a prudent approach is to start with a relatively simple model and add complexity as it is needed.

Modeling is critical to a complete response to question 4 (How do we preclude gaps and overages?) as well. The organization changes policies and practices to influence the characteristics of the future workforce. But exactly what effect will each change bring about, and what effect will the changes have in combination? Managers generally have little trouble identifying changes to policies and practices because most are attempts to overcome obstacles these managers face daily. But changes are not without costs (not only financial costs but also employees' resistance, their need for more education, and the initial decline in their productivity). Changes influence different characteristics (some, for example, may increase workforce experience; others may influence skills). Changes influence the same characteristic to differing degrees (some, for example, may have a dramatic effect on increasing experience; others a marginal effect). Changes may work together or be at odds with each other (some, such as bonuses tied to performance, may increase experience; others, such as variable separation incentive pay, may decrease experience). Estimating potential changes in terms of their effects on critical workforce characteristics can help the organization assign priorities—and thereby resource programs—effectively.

Again, although workforce planners may be able to evaluate the effects of individual changes without a model, estimating the effects of multiple, interacting changes requires one. In addition, workforce planners are unlikely to know the effects of changes with a high degree of confidence. A model can provide insight into the potential effects through such techniques as sensitivity analysis. Finally, a model forces the participants in the workforce planning activity to make their assumptions explicit—and thus positions them for debate—leading to a better shared understanding of how the business works. (Chapter Four elaborates on the last point.)

Although some organizations may have developed models capable of supporting workforce planning, customized for use with their civil service workforces, an organization just beginning to conduct workforce planning can expect to be disappointed with the availability of inventory projection models suitable to its needs. Appendix A describes the essential elements of a simple workforce projection model assembled by RAND, which is applicable within any organization.[10] Of course, an organization will also need to develop

---

[10]RAND assembled a simple workforce projection model for use by the defense acquisition community, populated it with relevant data, and made it available to

organization-specific supporting data in conjunction with the RAND or other models.

---

acquisition workforce planners in the DoD components and the major acquisition commands.

Chapter Four

A PROCESS VIEW: FOUR STEPS TO WORKFORCE PLANNING

The blueprint described in Chapter Three provides the foundation—the information needed—to answer the four thematic questions underlying workforce planning. This chapter describes a four-step process any organization can use to engage its key participants in gathering and employing that information to accomplish the three purposes of workforce planning set forth in Chapter Two. We begin by defining the intent of the process. We then describe the four steps. We conclude with observations on the key features of the process—related to the critical role of executive and line judgments, an incremental approach for using existing data to estimate desired distributions, and employing models to develop shared understanding among participants just discussed—in the context of how they support critical workforce planning activities.

GENERAL DESCRIPTION

The third purpose of workforce planning is to develop a convincing rationale for changes in policies and practices and for the reallocation of program resources to support the changes. Workforce planners have the task of articulating that convincing rationale—the business case that links the organization's strategic intent to those changes and reallocations. Figure 4.1 highlights the workforce planning process as the means for articulating that rationale.

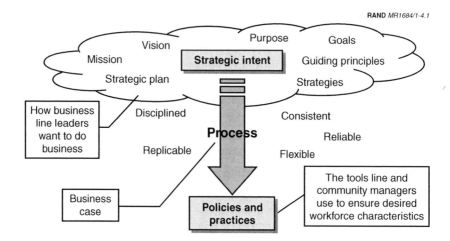

Figure 4.1—Articulating a Business Case for Change

The figure characterizes strategic intent as a broad statement of what business the organization is in and how the leaders of the organization want to carry out that business in the future. An organization generally captures its *strategic intent* in its vision, purpose, strategic plan, goals, etc. *Policies and practices* are the tools line managers and community managers use to influence hiring, development, retention, and separation—thereby shaping human capital to accomplish the organization's strategic intent.

The next subsection describes a *process* for engaging key participants—beginning with the organization's strategic intent and ending with identified changes to human resource management policies and practices—as the means for developing the business case. To provide a foundation acceptable within a federal agency, to other executive agencies, and to Congress, the process must be *consistent* (provide similar results in organizations with similar missions), *replicable* (provide similar results when repeated in an organization), and *reliable* (provide policies and practices that help achieve the strategic intent in the ways intended). To be applicable to all kinds of organizations, the process must be *disciplined* (structured to effectively use participants' time) and *flexible* (able to reflect the unique needs of different organizations).

A FOUR-STEP PROCESS

During the initial cycle of workforce planning in the DoD acquisition community (completed during the summer of 2001), the DoD components applied the blueprint described in Chapter Three. It became clear that much of the information needed to carry out workforce planning, even in the context of the blueprint, was available only at the business unit level. This was especially apparent with respect to articulating strategic intent in enough detail to identify critical workforce characteristics and their future desired distribution. The components generally recognized the need for greater input from executives and line managers in business units. We developed the four-step process portrayed in Figure 4.2 to focus the contributions of key participants at the business unit level.

The workforce planning process begins with the business unit's statement of its strategic intent. Separate methods exist for developing an organization's strategic intent (see, for example, Hax and Majluf, 1996; Hamel, 2000, particularly Chapter Three; and—more generally—Mintzberg, Ahlstrand, and Lampel, 1998). The first step of

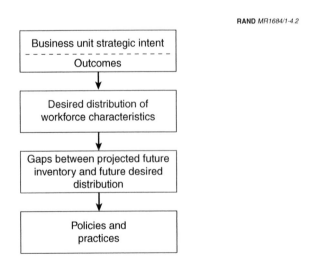

RAND *MR1684/1-4.2*

Figure 4.2—A Four-Step Process for Workforce Planning

the process assumes that an organization's strategic intent exists, usually expressed in the form of a strategic or business plan.[1] Some organizations have engaged in many cycles of strategic planning. However, none of the handful of plans we reviewed in detail contained language specific enough to identify implications for the workforce needed to carry out the strategic intent described in the plan. The first step of the workforce planning process, then, is to examine strategic intent in more depth to discover its implications for the workforce.[2]

It is also useful during the first step in the process to assess how the organization's environment has shifted or is expected to shift. Such shifts might explicitly influence an organization's formulation of its strategic intent. Alternatively, an organization might determine that an environmental shift (for example, an expected change in customer demand) does not alter its strategic intent but does alter the workforce needed to attain it.

Organizational outcomes can be a useful construct to enable an organization to translate statements of strategic intent into implications for the workforce. *Outcomes* are measurable results the organization produces that matter to its customers. Ideally, statements of the organization's strategic intent include desired outcomes; however, strategic plans and other expressions of that strategic intent generally do not explicitly delineate outcomes. Most often, they focus on *organizational actions* or *activities* rather than on the *results* the workforce is expected to produce. It is the results the organization produces for customers, however, that ultimately determine the resources required—in the case of workforce planning, the capabilities required in the workforce.[3] For example, an organizational outcome might be expressed as "capable, affordable, timely products reflecting an understanding of market/customer needs and requirements." This statement invites workforce planning participants to ask and

---

[1]This assumption is generally valid in DoD acquisition organizations; for example, we found no organization without some form of strategic intent.

[2]If an organization does not have a strategic plan or other expression of strategic intent, it must develop such a plan separate from the workforce planning process outlined here and before proceeding any further.

[3]The actions or activities the organization commits to carry out the strategic plan also require resources, but that is a different issue.

answer the questions, What kind of workforce is needed to ensure capable products? Is a different kind of workforce needed to ensure affordable or timely products? What characteristics (e.g., occupation, education, experience) should the workforce have to understand customer needs?[4]

Capturing the answers to such questions in terms of workforce characteristics is part of the second step of the workforce planning process. As defined earlier, a workforce characteristic is a concrete, measurable aspect of a group of workers that human resource management policy decisions can influence and that is critical for organizational success. The future desired workforce distribution might reflect a changed mix of occupations (for example, fewer mechanical engineers and logisticians and more information technologists and program managers) than in the current desired distribution. One should not be misled by the brevity of the description of the step(s). The participants in the process will not easily answer questions such as, What characteristics should the workforce possess to understand customer needs? First, they will need to address which workforce characteristics are relevant: Occupation? Educational level? Length of a particular kind of experience, such as acquisition experience? Second, if length of acquisition experience is critical, how much is enough? These are tough questions to answer; the process simply frames the questions. The answers lie within the participants, not the process.[5]

The third step of the workforce planning process is a traditional *gap analysis*. In this step, workforce planners identify the estimated over- or undersupply of people with critical characteristics in the workforce (the projected future inventory) compared with the future desired distribution. This relatively mechanical step depends critically on the ability to project the current inventory based on continuation of current policies and practices and on the product of the second step—an articulation of the workforce needed to accomplish the or-

---

[4]The participants may find that links between outcomes (desired results) and inputs (desired workforce characteristics) are not readily apparent. They may need to identify and assess intervening variables, such as outputs (products and services) used to generate desired outcomes or processes for producing these outputs. We offer some thoughts on applying these additional considerations later in the report.

[5]Later in the report, we describe an incremental process to assist these deliberations.

ganization's strategic intent (the future desired distribution). The overages indicate the need for human resource management policies and practices to retrain or separate those, for example, in occupations no longer required in the organization or required to a lesser degree than projected to be available (assuming continuation of current policies and practices). Shortages indicate the need for policies and practices to increase hiring, to motivate the individuals with critical characteristics to continue employment at higher rates than expected, and to retrain employees from occupations in oversupply. The gaps help target policies and practices to the relevant population, avoiding unnecessary resource expenditures and tailoring the most effective argument for change.

In the final step, participants identify specific policies and practices that will acquire, assign, develop, assess, motivate, reward, and separate people as necessary to close the gaps. The participants should describe the policies and practices with sufficient specificity for a human resource management specialist to flesh them out in detail. In addition, this step is the appropriate place for workforce planners to estimate the potential effect of the identified policies and practices on reducing the gaps identified in the third step. Further, the participants must determine which policies and practices will require new or modified authority and resources for implementation. Together, the estimate of the potential effects and an assessment of the difficulty of acquiring authority and resources serve as the basis for prioritization and a plan of action for proceeding.

Appendix B contains a sample agenda for conducting the workforce planning process.

AN ITERATIVE PROCESS

Although we present the process as unfolding linearly in Figure 4.2, its application is most effective when conducted iteratively. For example, while identifying critical characteristics in the second step, the participants may find it valuable to revisit and restate the outcomes or add new outcomes in the first step. Similarly, the second, third, and fourth steps benefit from interaction. The left side of Figure 4.3 presents the workforce planning process as an organization might most effectively execute it.

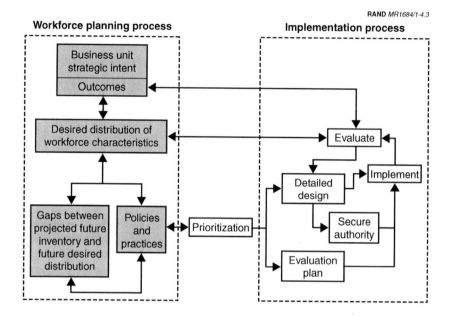

Figure 4.3—An Iterative Process

Completing the four steps of the workforce planning process is not the end of the effort. The policies and practices the workforce planning participants identify will vary in terms of cost, likelihood of having the intended effect, employee reaction, managerial competence (and willingness) to apply, difficulty in obtaining authority to implement, the resources of the human resource management staff to support, and so forth. The organization will undoubtedly want to prioritize the policies and practices identified and to develop an action plan for implementation.

The right side of Figure 4.3 portrays the relationship between the workforce planning process and the follow-on implementation process required to ensure that the effort devoted to workforce planning is fruitful. Detailed design of the policies and practices, development of an evaluation plan (to measure the actual effect of the changes) and, in some cases, securing the appropriate authority all precede implementation. An evaluation step ensures feedback.

KEY FEATURES OF THE FOUR-STEP PROCESS

The process outlined in the preceding sections is a general guide to help focus the contributions of participants on some of the most difficult, but important, activities of workforce planning. It does not provide a mechanistic procedure leading to specific answers. Rather, it is a structure within which leaders in the organization can engage in a facilitated dialogue.[6]

Although we do not argue that facilitated dialogue is the *only* means of carrying out the process, based on our experience (and as part of a similar process focused on organizational behavior), we believe it produces a synergy not available through a mechanistic procedure. In particular, the workforce planning process relies heavily not just on human resource management expertise, but on unique and critical expertise and perspectives that individual executives, line managers, community managers, and staff principals bring to the process. For example, executives understand and usually talk about strategic intent in a language nearly devoid of human capital implications; line managers appropriately are narrowly focused on their ability to meet near-term objectives; community managers and human resource managers often have only a cursory understanding of how they can help the organization succeed. Facilitated dialogue is a method for carrying on a conversation during which all participants learn, understand, appreciate, and build upon the others' perspectives. Such a conversation is key to successfully linking strategic intent to human resource management policies and practices. A facilitated dialogue is not appropriate for all organizations, but for those that find it useful, the workforce planning process provides an application.[7]

---

[6]We use the term *facilitated dialogue* in the same sense as Peter M. Senge when discussing team learning (1990, Chapter 12, p. 233 ff) and more generally by Ellinor and Gerard (1998).

[7]Facilitated dialogue also has potential benefits beyond workforce planning. Although the diversity of viewpoints the executives, line managers, community managers, and staff principals bring to the table enriches the product of workforce planning, it raises many issues workforce planning cannot resolve, thereby developing a greater appreciation of the value of other aspects of human capital strategic planning (cultural, organizational, and behavioral components). In addition, it has helped some organizations we have worked with develop a shared understanding and common focus among the participants regarding what the leaders want to accomplish and how they want to ac-

Other than facilitated dialogue, three features make the workforce planning process distinctly different from processes other practitioners suggest: focusing on strategic intent, estimating future desired distributions incrementally, and using inventory projection models to assist workforce planning participants in quantifying the expected impact of their recommendations.

These three features correspond to the critical considerations described in Chapter Three (executive and line judgments, data, and modeling); however, this chapter describes them in terms of how the organization can focus the contributions of the participants to address these considerations.

Assessing Strategic Intent

The heart of the business case for changes to policies and practices lies in the assessment of strategic intent. This is the step practitioners of workforce planning most often praise and then neglect. A variety of frameworks can be brought to bear to help an organization think through the link between strategic intent and the workforce characteristics needed to carry it out.

We believe that participants' focus on desired organizational outcomes can bring strategic intent explicitly into the workforce planning process. The focus on desired outcomes instills value in two important ways. First, it brings the customer's perspective to the fore. Although this perspective is often included in the development of a strategic plan, it less often influences human capital decisions. Second, it highlights *measurable* results the organization produces. It thereby forms the basis for eventually evaluating the effectiveness of changes to policies and practices flowing from the workforce planning process and ultimately for weaving accountability into the performance management system. If the organization's leaders are to value the workforce planning process, it must positively affect the results that matter to the organization.

---

complish it. Finally, we have found that the workforce planning "dialogue" often stimulates potential improvements in the organization's overall strategic planning process because senior leaders value its nontraditional approach.

Desired outcomes provide a focus for the workforce planning participants to convert their strategic plan into workforce characteristics. The leaders of some organizations may find engaging with other leaders in facilitated dialogue to discern desired outcomes such a valuable exercise that they decide to adopt this process as an ongoing strategic planning activity.[8] However, leaders of other organizations may want to probe beyond the identification of desired outcomes—even on their initial foray into workforce planning. In any event, greater specificity is desirable as workforce planning matures. To that end, we offer some avenues an organization can pursue beyond outcomes.

Desired outcomes may suggest or highlight broad changes—for example, in an organization's business model, in the value propositions it intends to offer its customers, or in the organizational core competencies it seeks to develop. Although discussion of these strategic concepts may be nearly as general as outcomes themselves, an organization may have adopted one of these concepts and developed it enough to provide a clearer line of sight between the organization's strategic intent and the workforce characteristics needed to accomplish it.[9] Indeed, while we see the concept of outcomes as useful in view of its central position in the Government Performance and Results Act of 1993 (and related government-wide strategic initiatives), business models, value propositions, or core competencies could provide a viable alternative starting point for identifying and assessing critical workforce characteristics if one of those views is fully developed in the organization.

Another, more traditional framework may be useful in other organizations. Such a framework might start with desired outcomes and ask, in turn, What organizational outputs (products and services) are

---

[8]During the initial application of workforce planning, one Department of the Navy (DON) business unit employed the workforce planning process with a major focus on outcomes as the means to highlight the interdependencies of its business lines and the community managers.

[9]One DON business unit had developed organizational core competencies; the other business unit was in the process of considering a fundamental change in its business model. RAND has helped a third business unit outside the context of workforce planning develop a wide ranging set of "core equities," similar in many regards to organizational core competencies; the line of sight between the equities and workforce characteristics is, in many instances, quite direct. (See Hynes et. al., 2002.)

needed to accomplish the outcomes? What technologies, organizational systems, and processes are needed to produce and deliver the outputs? What activities/tasks are needed to employ the technologies and systems and to carry out the processes? and What resources (inputs) are needed to carry out the activities? Such a sequence of questions leads to more-detailed information for discerning critical workforce characteristics and their distribution (inputs).

Different organizations, at different stages of maturity in workforce planning, will find some frameworks better suited to their needs than others. Also, the rigor of assessment of the link between strategic intent and workforce characteristics (and, correspondingly, the comprehensiveness of the recommendations resulting from the assessment) should not exceed (at least by too much) the availability of the data or the ability of models to make use of them.

Estimating Future Desired Distributions Incrementally

Participants find the second step of the workforce planning process particularly challenging. However, estimating the future desired distribution of critical workforce characteristics and elucidating the link between the future desired distribution and the desired outcomes are critically important to the business case for changing policies and practices. Therefore, the participants must confidently complete this step to ensure an effective result.

A thorough understanding of an organization's environment and a clear statement of its desired outcomes (supplemented with additional information such as outputs and processes) should provide the major source of information needed to identify critical workforce characteristics and to estimate their future desired distribution. Ideally, based on the organization's desired outcomes and expected environment, workforce planners assess the future workload of the organization (in terms of level and type of work, for example) and employ a workload model (which helps translate expected workloads into requirements for workers) to estimate the future desired distribution of the critical workforce characteristics. Practically, this *synoptic* approach, illustrated in Figure 4.4, may be too much to expect, however, especially during the early application of workforce planning in an organization. Consequently, in conducting the workforce planning process, we recommend an *incremental*

RAND *MR1484/1-4.4*

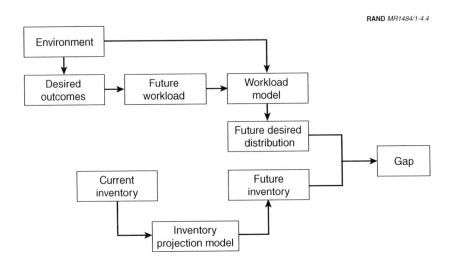

**Figure 4.4—A Synoptic Approach for Determining the Future Desired
Distribution**

approach to assist the workforce planning participants in estimating
the future desired distribution, as illustrated in Figure 4.5.[10]

The incremental approach begins with the current inventory.
Business unit executives and line managers review the information
regarding the distribution of the critical workforce characteristics
within the current workforce (current inventory). These participants
perform two assessments that lead to an estimate of the current desired distribution of workforce characteristics. First, they assess
whether their environment has changed (D_1) in ways that should
have influenced the current composition of the workforce but

---

[10]We use the terms "synoptic" and "incremental" in the senses conveyed by Charles
E. Lindblom in his 1959 and 1979 articles in *Public Administration Review* on the in-
evitability of partial rather than complete analysis of complex public policy alterna-
tives. A *synoptic* approach, in our context, would be characterized by a comprehensive
understanding of all workload and other related factors that influence the desired
distribution of workforce characteristics, enabling workforce planners to model the
relationships between workloads and workforces. Lacking such comprehensive
knowledge, planners can make strategic improvements in their workforces through
successive, *incremental* departures from the status quo, basing the direction and
magnitude of each incremental departure on whatever subjective or objective
indicators are available to them.

heretofore have not. Second, they assess whether the outcomes they are currently producing differ in important ways from the current desired outcomes (D_2). If the participants identify no changes between the past and current environment or between current actual and desired outcomes, the current inventory becomes the estimate of the current desired distribution. If they identify changes between the past and current environment or between current actual and desired outcomes $(D_1$ or $D_2)$, the participants estimate the changes to the distribution of critical characteristics in the current inventory (D_3) needed to bring their workforce in line with the demands of their current environment and their desired outcomes. The result of this exercise is an estimate of the current desired distribution of critical workforce characteristics. The left side of Figure 4.5 portrays the first step of the incremental approach.

The participants then focus on the future environment and future desired outcomes to develop a second increment of changes in the

RAND MR1484/1-4.5

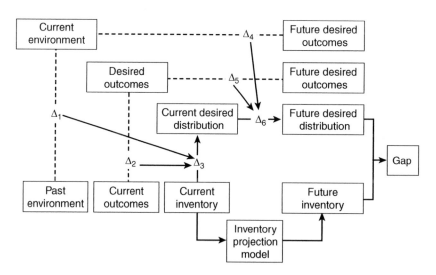

Figure 4.5—An Incremental Approach for Determining the Future
Desired Distribution

desired distribution of workforce characteristics. In this increment, business unit executives and line managers perform two additional assessments within the span of the planning horizon. First, they identify expected changes between the current and future environment that will affect the desired composition of the workforce. Second, they identify the changes between current desired outcomes and future desired outcomes. If they identify changes in the environment or desired outcomes (D_4 or D_5), the participants estimate the changes in the distribution of workforce characteristics (D_6) needed to bring the workforce in line with the future environment and future desired outcomes, resulting in a future desired distribution of workforce characteristics that differs from the current desired distribution. If no environmental or desired outcome changes are expected, the current desired distribution of the workforce is also the future desired distribution.

Modeling Future Inventory

Most participants in the workforce planning process can intuit the effect of a particular practice (say, that a bonus tied to clear individual or group performance metrics will result in greater retention as well as greater performance). Their intuition, however, is unlikely to yield a quantitative estimate of the impact on retention, and even if it does, the estimates of different participants will almost certainly vary. Quantification and consensus are important in prioritizing the results of workforce planning because the environment and the workforce are in a constant state of flux and resources (program and staff) to implement changes to policies and practices are generally limited.

In particular, the degree to which the organization implements a new practice (for example, how broadly to apply a hiring bonus and the size of that bonus) will depend on a number of factors. These factors include, for example, the size of an expected shortage in a critical workforce characteristic (which, in turn, depends on the estimated effect of long-standing and recently implemented policies and practices in the face of a changing environment) and the expected impact of other policies and practices (for example, an expedited hiring process).

If the organization is considering an array of policies and practices to address a variety of workforce issues, the effective expenditure of resources demands that the participants prioritize and tailor the policies and practices to reflect their interactive effects. Of course, the impact of an individual policy or practice can only be known when it is implemented and its effect measured. However, participants can (and must) estimate the expected impact in order to make informed decisions. An inventory projection model (which depicts how the expected composition of a workforce will change over time) is essential to, and can facilitate, bringing the potentially diverse judgments of workforce planning participants to bear and developing their shared understanding of the expected impact of their recommendations.

An inventory projection model requires participants to make implicit assumptions explicit. For example, in its simplest form, a model may start with the current total workforce distributed by years of service. It may project that inventory forward using hiring rates and loss rates for aggregate individuals in each year of service based on past gains and losses. Even in this simple case, workforce planners face serious questions. Are past rates appropriate for the future? Were policies in place in the past (say, during a drawdown) that affected these rates, but will not be present in the future? Has the organization implemented recent changes in policies and practices, the effects of which are not yet fully visible in past rates? If so, how should past rates be adjusted to reflect a likely estimate for the future? Again, the reason for asking such questions and engaging in dialogue to answer them is not to come up with a sterile "right" answer. Rather, it is to engage the participants, obtaining their best estimate of the effects, making assumptions visible, and thereby allowing them to test the accuracy of their assumptions over time and to revise them accordingly.

Different assumptions regarding gains and losses will affect the future inventory and, in turn, any gaps between it and the future desired distribution. The nature of the gaps influences the choice and design of changes to policies and practices. But this is only one side of the coin. Potential changes to policies and practices also have an effect on hiring and loss rates. The number and magnitude of changes needed to close the gap depend on their effect on aggregate rates as manifested through individual policies and practices. Using the inventory projection model to project the estimated effect of recommended policies and practices gives participants the opportunity to make their intuition explicit.

In effect, the assumptions about effects largely drive the selection of policies and practices, and participants should make their assumptions explicit and argue about them when evaluating which policy or practice is more desirable. The inventory projection model affords a forum for this discussion.

We have shown the importance of executive and line participation, data, and modeling—both in answering the four thematic questions and in focusing the contributions of the participants in workforce planning. In the next chapter, we present recommendations for enhancing those key success factors.

Chapter Five

KEY SUCCESS FACTORS IN WORKFORCE PLANNING

As described in Chapter Three, three key factors lead to successful workforce planning: active executive and line participation, accurate and relevant data, and sophisticated workload and inventory projection models. In this chapter, we set out our major recommendations relevant to these factors to enhance the effectiveness of the workforce planning activity. We end with a concluding summary.

ACTIVE EXECUTIVE AND LINE PARTICIPATION

The line of sight between how leaders want to carry out business in the future and the human capital capabilities needed to do that effectively is clearest at the business unit level (or below). Even so, characterizing the capabilities needed is not an easy task. This aspect of workforce planning benefits from the active participation of business unit executives (who determine how they intend to carry out their business) and line managers (who determine the capabilities needed).

Leaders at levels above the business unit also play an important, but different, role in the workforce planning activity. Their role is critical. They

- translate higher-level direction into clear guidance for line organizations

- integrate the results across business units and larger organizational entities and tailor the business case for change for the higher levels

- support the results of workforce planning at the lower levels.

We recommend that senior leaders (above the business unit level) explicitly delineate the roles and responsibilities of executives throughout the organization involved in the workforce planning activity. In particular, we recommend that these senior leaders encourage and reward active participation in the workforce planning activity by business unit executives and line managers.

ACCURATE AND RELEVANT DATA

Data on workforce characteristics are the common language of workforce planning. Although much of the workforce planning activity is best carried out by individual business units, the kind of data needed is similar throughout the organization (across business units), suggesting that the workforce planning activity benefits from a common source of data.

We recommend that the organization's corporate headquarters take the lead to develop the functional specifications for a human resource information system to support workforce planning, both in its near-term application and as workforce planning matures.

SOPHISTICATED WORKLOAD AND INVENTORY PROJECTION MODELS

Insight into how the composition of the workforce may change over time (with or without changes in the environment or in organizational policies) informs human capital decisions. Data are the basis of these estimates; models are the means. As with data, the kind of models needed may be similar throughout the organization (although the level of sophistication necessary may differ among business units). However, models can be expensive to develop and maintain.

We recommend that the organization's corporate headquarters evaluate the availability, costs, and benefits of comprehensive, integrated workload and inventory projection models for all categories of employees (including contractors if contractor personnel are integrated with civil service employees). If this evaluation suggests a positive return, we further recommend that the organization's corporate head-quarters invest in the development of such models, which business units would modify for their individual situations.

CONCLUDING SUMMARY

This report describes a process any organization can use to conduct workforce planning. We have examined workforce planning from three points of view, to help an organization's leaders

- decide whether they want to engage in workforce planning (a goal-oriented view)

- understand the mechanics of workforce planning and identify the resources needed (a structural view)

- mobilize the key participants to carry out workforce planning (a process view).

In sum, we intend this report to serve as a user's guide for partici-pants conducting workforce planning as they begin to institute the activity in their organization.

A companion report, *An Executive Perspective on Workforce Planning*, MR-1684/2-OSD, completes the context from a uniquely executive vantage point. Together, the two reports fully describe the initial application of workforce planning in the DoD acquisition community, completed during the summer of 2001.

A. THE ACQUISITION WORKFORCE PROJECTION MODEL

A simple acquisition workforce projection model, developed by RAND, was made available to acquisition workforce planners during the 2001 and 2002 acquisition workforce planning cycles in the form of a Microsoft Excel workbook customized for this purpose. RAND designed this model to estimate how longevity-related characteristics of a workforce are likely to change over a period of several years. The model also enables workforce planners to estimate how alternative human resource programs or policies might affect future inventories of critical characteristics.

The basic workforce characteristic depicted in the model is *year of service* (YOS). The model accepts, as user input, the *beginning inventory* of a workforce, distributed from YOS 1 to YOS 50.[1] The user must also supply *target end strengths,* i.e., the planned total size of the workforce at the end of the current and each future fiscal year. The model uses *continuation rates* to calculate the number of workers in each YOS who are expected to remain in the workforce for an additional year. Workers who do not continue from one year to the next are counted as *losses.* The model then determines the number of *gains* (new hires) necessary to replace losses while accounting for any change in the target end strength. Although most gains are in YOS 1, many have previous federal service and therefore enter the

---

[1]Although a workforce may contain a few workers with over 50 years of service, their number will be small compared to the total. They can be ignored without significantly affecting model results.

workforce in YOS 2 and above. In summary, the model starts with a workforce as it looks at the end of the most recent fiscal year and depicts how it might look at the end of each successive fiscal year.

Figure A.1 illustrates the basic configuration of the model. Column A of the spreadsheet indicates the YOS. Column B contains the beginning inventory as it looked at the end of FY 2001.[2] Columns C through L contain the projected workforce at the ends of fiscal years 2002 through 2011. Line 56 contains the target end strengths for each fiscal year (set, in this example, to maintain an unchanging workforce size of 115,000 during the projection period). As would be expected, the YOS populations generally get smaller as YOS increases. However, this is not always true. Populations in YOS 11 and below are smaller than populations in YOS 12 through 35—a reflection of low hiring levels during the workforce drawdown of the past decade.

Figure A.2 shows the bottom half (rows 54 through 74) of the same sheet depicted in Figure A.1. These rows provide summary statistics regarding the workforce. These include total strengths, total losses and gains, net losses and gains, turnover rates, and several aggregate measures of experience. Note, for example, that the average YOS drops from 20.3 years in FY 2001 to 18.3 years by FY 2011. This expected rejuvenation of the workforce can also be seen in the rows that depict proportions of the workforce in various YOS groupings. Over time, the less experienced proportions of the workforce in YOS 1–3 and YOS 4–10 continually increase while the more experienced proportions in YOS 11–20 decline. Additionally, the last three lines provide an estimate of how much of the current (end FY 2001) workforce will have retired or separated by the end of each projected

---

[2]The data used in this example reflect the beginning inventory embedded in the workbook when it was initially supplied to users. This beginning inventory depicts the DoD-wide acquisition workforce, counted using the "Packard algorithm." This algorithm is a modified version of an approach for counting the defense acquisition workforce developed by the 1986 President's Blue Ribbon Commission on Defense Management (the Packard Commission). See, for example, Burman, Cavallini, and Harris (1999) or information contained at http://www.acq.osd.mil/dpap/workforce/careermanagement, accessed on November 13, 2003.

RAND*MR1684/1-A.1*

Projected Inventory - Base Case

YOS	Beginning Inventory (End of FY 2001)	2002	2003	2004	2005	2006	2007	2008	2009	2010	2011
1	2938	4160	4449	4624	4758	4872	4988	5083	5159	5245	5312
2	2140	2870	4003	4276	4441	4569	4679	4790	4880	4953	5035
3	1419	2102	2788	3843	4100	4257	4379	4483	4588	4674	4744
4	1043	1443	2084	2723	3702	3943	4092	4207	4307	4406	4488
5	1143	1190	1571	2166	2757	3656	3883	4024	4134	4229	4324
6	1242	1205	1258	1619	2176	2729	3570	3784	3917	4022	4113
7	1095	1285	1258	1313	1655	2180	2701	3491	3693	3821	3921
8	916	1116	1300	1279	1333	1658	2155	2649	3395	3587	3709
9	958	959	1154	1331	1314	1368	1676	2146	2612	3316	3499
10	1349	1009	1016	1206	1378	1365	1419	1714	2164	2609	3281
11	2009	1383	1071	1083	1265	1430	1420	1474	1754	2179	2601
12	2399	2020	1433	1140	1155	1331	1492	1485	1538	1806	2212
13	5075	2394	2040	1485	1210	1227	1398	1553	1549	1602	1859
14	3338	4962	2379	2040	1506	1242	1260	1426	1577	1575	1626
15	4743	3275	4841	2360	2036	1525	1273	1292	1453	1600	1598
16	4393	4634	3226	4736	2350	2040	1549	1308	1328	1484	1627
17	6518	4318	4556	3193	4659	2346	2047	1573	1340	1360	1513
18	5926	6342	4222	4454	3140	4557	2326	2039	1582	1358	1379
19	5264	5738	6140	4103	4328	3064	4428	2283	2008	1569	1354
20	5197	5111	5570	5959	3997	4214	2998	4313	2247	1982	1560
21	5319	5026	4944	5387	5763	3876	4085	2916	4182	2194	1940
22	4930	5125	4846	4769	5195	5555	3749	3951	2831	4044	2141
23	4475	4778	4968	4701	4628	5038	5386	3648	3842	2765	3933
24	4386	4355	4650	4835	4577	4507	4905	5242	3559	3748	2705
25	3764	4252	4223	4509	4688	4439	4371	4756	5083	3455	3637
26	3291	3633	4103	4075	4351	4523	4284	4220	4590	4904	3340
27	3527	3176	3505	3958	3932	4197	4363	4133	4072	4428	4730
28	4219	3408	3070	3388	3824	3799	4055	4215	3994	3935	4278
29	3385	4084	3302	2977	3284	3705	3682	3929	4084	3871	3814
30	3032	3267	3941	3188	2874	3170	3577	3554	3792	3942	3736
31	2972	2803	3020	3643	2947	2658	2931	3307	3286	3506	3644
32	2318	2700	2547	2745	3309	2678	2416	2664	3005	2986	3186
33	2281	2120	2470	2330	2510	3026	2450	2210	2437	2748	2731
34	2169	2000	1859	2166	2043	2201	2654	2148	1938	2137	2410
35	1885	1886	1739	1616	1883	1776	1914	2307	1868	1685	1858
36	1435	1610	1611	1486	1381	1609	1518	1635	1971	1596	1440
37	741	1166	1308	1309	1207	1122	1307	1233	1328	1601	1297
38	569	576	906	1017	1017	938	872	1016	958	1033	1245
39	416	439	445	700	785	786	725	674	785	740	797
40	334	314	332	336	528	593	593	547	508	532	559
41	233	255	239	253	256	403	452	452	417	388	451
42	135	167	183	172	181	184	289	324	324	299	278
43	80	89	111	121	114	120	122	191	215	215	198
44	61	80	89	111	121	114	120	122	191	215	215
45	53	52	68	76	94	103	97	102	104	163	183
46	27	53	52	68	76	94	103	97	102	104	163
47	25	27	53	52	68	76	94	103	97	102	104
48	14	19	21	41	40	53	59	73	80	75	79
49	10	14	19	21	41	40	53	59	73	80	75
50	50	10	14	19	21	41	40	53	59	73	80
Total	115241	115000	115000	115000	115000	115000	115000	115000	115000	115000	115000
Target end strength		115000	115000	115000	115000	115000	115000	115000	115000	115000	115000

Base Model / Base Rates / Alt Model / Alt Rates / Comparison / DMDC loss data FY00 / DMDC

Figure A.1—Basic Configuration of the Workforce Projection Model

RANDMR1684/1-A.2

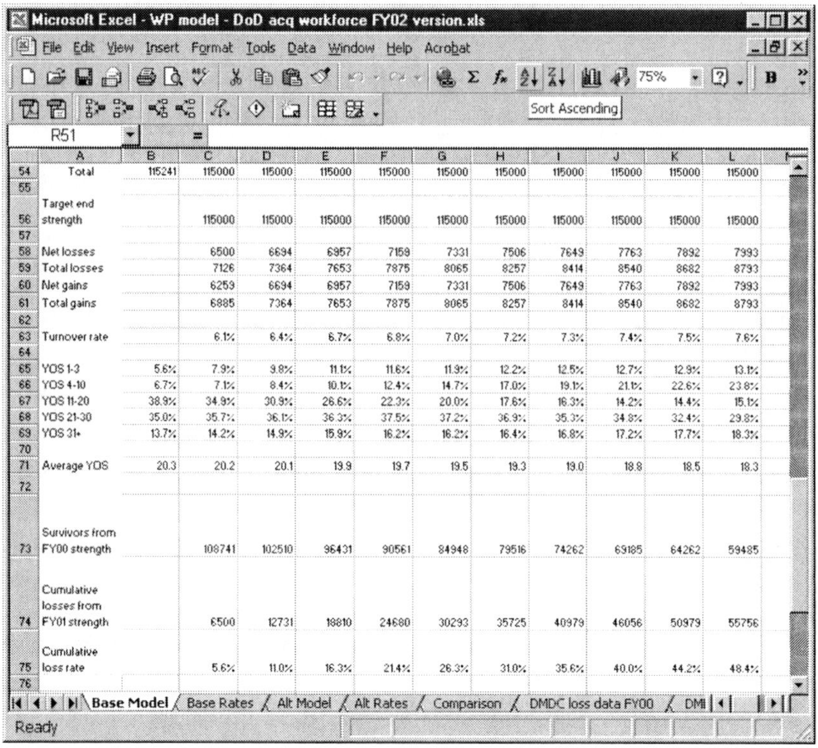

	A	B	C	D	E	F	G	H	I	J	K	L
54	Total	115241	115000	115000	115000	115000	115000	115000	115000	115000	115000	115000
55												
56	Target end strength		115000	115000	115000	115000	115000	115000	115000	115000	115000	115000
57												
58	Net losses		6500	6694	6957	7159	7331	7506	7649	7763	7892	7993
59	Total losses		7126	7364	7653	7875	8065	8257	8414	8540	8682	8793
60	Net gains		6259	6694	6957	7159	7331	7506	7649	7763	7892	7993
61	Total gains		6885	7364	7653	7875	8065	8257	8414	8540	8682	8793
62												
63	Turnover rate		6.1%	6.4%	6.7%	6.8%	7.0%	7.2%	7.3%	7.4%	7.5%	7.6%
64												
65	YOS 1-3	5.6%	7.9%	9.8%	11.1%	11.6%	11.9%	12.2%	12.5%	12.7%	12.9%	13.1%
66	YOS 4-10	6.7%	7.1%	8.4%	10.1%	12.4%	14.7%	17.0%	19.1%	21.1%	22.6%	23.8%
67	YOS 11-20	38.9%	34.9%	30.9%	26.6%	22.3%	20.0%	17.6%	16.3%	14.2%	14.4%	15.1%
68	YOS 21-30	35.0%	35.7%	36.1%	36.3%	37.5%	37.2%	36.9%	35.3%	34.8%	32.4%	29.8%
69	YOS 31+	13.7%	14.2%	14.9%	15.9%	16.2%	16.2%	16.4%	16.8%	17.2%	17.7%	18.3%
70												
71	Average YOS	20.3	20.2	20.1	19.9	19.7	19.5	19.3	19.0	18.8	18.5	18.3
72												
73	Survivors from FY00 strength		108741	102510	96431	90561	84948	79516	74262	69185	64262	59485
74	Cumulative losses from FY01 strength		6500	12731	18810	24680	30293	35725	40979	46056	50979	55756
75	Cumulative loss rate		5.6%	11.0%	16.3%	21.4%	26.3%	31.0%	35.6%	40.0%	44.2%	48.4%
76												

Figure A.2—Summary Statistics in the Workforce Projection Model

fiscal year. These data suggest that almost half (48.4 percent) of the current workforce will have departed by FY 2011.

The spreadsheet depicted in Figures A.1 and A.2 contains formulas that calculate the expected workforce inventory in each successive fiscal year. Each year group progresses through the matrix on a diagonal path. For example, the beginning inventory has 4,743 workers in YOS 15. About 97 percent of these workers continue on to YOS 16 in FY 2002. They are joined in YOS 16 by about 1 percent of the 6,885 new hires in FY 2002 (those who enter with 15 years of previous service), resulting in a YOS 16 population of 4,634 by the end of FY 2002. These 4,634 workers are similarly "aged" to arrive at the YOS 17 population of 4,556 by the end of FY 2003. (See the shaded diagonal in the close-up view of the spreadsheet in Figure A.3).

The formulas contained in the workforce sheet use continuation rates and the distribution of gains (new hires) by year of service. These rates are contained in a separate sheet in the workbook. Note, in Figures A.1 through A.3, that sheets in an Excel workbook are identified by a row of tabs along the bottom of the workbook window. The material depicted in Figures A.1 through A.3 is in a sheet labeled "Base Model." The accompanying rates are in a sheet labeled "Base Rates."

RANDMR1684/1-A.3

Microsoft Excel - WP model - DoD acq workforce FY02 version temp.xls

File Edit View Insert Format Tools Data Window Help Acrobat

L27   =K26*'Base Rates'!L27+L$60*'Base Rates'!L89

Projected Inventory - Base Case

YOS	Beginning Inventory (End of FY 2001)	2002	2003	2004	2005	2006	2007	2008	2009	2010	2011
1	2938	4160	4449	4624	4758	4872	4988	5083	5153	5245	5312
2	2140	2870	4003	4276	4441	4569	4679	4790	4880	4953	5035
3	1419	2102	2788	3843	4100	4257	4379	4483	4588	4674	4744
4	1043	1443	2084	2723	3702	3943	4032	4207	4307	4406	4488
5	1143	1190	1571	2166	2757	3656	3883	4024	4134	4223	4324
6	1242	1205	1258	1619	2176	2729	3570	3784	3917	4022	4113
7	1035	1285	1258	1313	1655	2180	2701	3491	3693	3821	3921
8	916	1116	1300	1279	1333	1658	2155	2649	3395	3587	3709
9	958	953	1154	1331	1314	1368	1676	2146	2612	3316	3439
10	1349	1009	1016	1206	1378	1365	1419	1714	2164	2609	3281
11	2009	1383	1071	1083	1265	1430	1420	1474	1754	2179	2601
12	2399	2020	1433	1140	1155	1331	1492	1485	1538	1806	2212
13	5075	2394	2040	1485	1210	1227	1398	1553	1549	1602	1859
14	3338	4962	2379	2040	1506	1242	1260	1426	1577	1575	1626
15	4743	3275	4841	2360	2036	1525	1273	1292	1453	1600	1598
16	4393	4634	3226	4736	2350	2040	1549	1308	1328	1484	1627
17	6518	4318	4556	3193	4659	2346	2047	1573	1340	1360	1513
18	5326	6342	4222	4454	3140	4557	2326	2039	1582	1358	1379
19	5264	5738	6140	4103	4328	3064	4428	2283	2008	1569	1354
20	5197	5111	5570	5959	3997	4214	2938	4313	2247	1982	1560
21	5319	5026	4944	5387	5763	3876	4085	2916	4182	2194	1940
22	4930	5125	4846	4769	5195	5555	3749	3951	2831	4044	2141
23	4475	4778	4968	4701	4628	5038	5386	3648	3842	2765	3933
24	4386	4355	4650	4835	4577	4507	4905	5242	3559	3748	2705
25	3764	4252	4223	4509	4688	4439	4371	4756	5083	3455	3637
26	3291	3633	4103	4075	4351	4523	4284	4220	4590	4904	3340
27	3527	3176	3505	3958	3932	4197	4363	4133	4072	4428	4730

Base Model / Base Rates / Alt Model / Alt Rates / Comparison / DMDC loss data FY00 / DM

Ready

Figure A.3—Diagonal Progression of a Year Group Through the Workforce Projection Model

B. SAMPLE AGENDA FOR CONDUCTING THE WORKFORCE PLANNING PROCESS

This sample agenda describes the objective of an application of the workforce planning process at a business unit, the appropriate workforce planning participants, and an outline of the sessions devoted to the process.

Objectives: In order to accomplish the organization's strategic intent:

1. Identify critical workforce characteristics (today and in the future).

2. Assess the desired distribution of people (today and in the future) within those critical workforce characteristics.

3. Develop policies and practices to acquire and shape the organization's workforce to reflect the desired distribution.

4. Agree on a plan for securing the authorities and resources needed to implement the policies and programs.

Secondary objective: Develop the capability and confidence of the participants to apply the workforce planning process in support of their individual business line's plans, as necessary, as well as on an ongoing basis.

Participants: Senior leadership and staff principals (including community managers and human resources), business line leaders, human resource management staff.

Session I (One day): Workforce planning process introduction; assessments of organization's strategic and business plans.

- Purpose: Develop and ensure a shared understanding of the workforce planning process and of the organization's strategic intent as reflected in planning documents; agree on the elements of workforce planning that lie in the purview of the organization's corporate headquarters and in the purview of the business lines; identify desired outcomes.

- Process: Participants review planning documents before the meeting; facilitator presents workforce planning framework; facilitator prompts dialogue among participants on the workforce planning process, the purview of workforce planning at the corporate and business line levels, and the elements of strategic intent on which the participants will focus for workforce planning.

- Result: Participants express willingness to proceed with the workforce planning process, agree on the elements of workforce planning most appropriately carried out at the corporate level and at the business line level, and identify desired organizational outcomes (and additional perspectives—such as products, processes, etc., required by the strategic intent) with enough specificity to elicit desired workforce characteristics.

Session II (Half day): Critical workforce characteristics.

- Purpose: Using the structured workforce planning process described in this document, identify workforce characteristics critical to the organization in accomplishing its strategic intent.

- Process: By means of facilitated dialogue among participants, elicit workforce characteristics critical to accomplishing corporate strategic intent using the desired outcomes as the primary rationale.

- Result: Agree on a complete set of workforce characteristics critical to accomplishing the organization's corporate strategic intent.

Session III (One day): Desired distribution of people within each critical workforce characteristic.

- Purpose: Using this structured workforce planning process, estimate the desired distribution of people within each critical workforce characteristic to accomplish the organization's strategic intent.

- Process: The organization presents current distribution of employees (inventory) for the workforce characteristics agreed upon in Session II (including appropriate cross-tabulations); facilitated dialogue among participants determines the current and future desired distribution for each critical characteristic. For each critical workforce characteristic, the participants start with the distribution of the current inventory and either (a) validate that the current actual distribution (inventory) adequately reflects the current desired distribution or (b) make incremental changes from the current actual distribution to specify the current desired distribution. Participants then estimate the future desired distribution by making judgments regarding the incremental effect of the environment, corporate and functional guidance, and the business unit's strategic intent on the agreed-upon current desired distribution.

- Result: Agreement on the future desired distribution for each critical workforce characteristic (or cross tabulations).

Session IV (Two days): Policies and practices to shape the workforce to the future desired distribution.

- Purpose: Identify the policies and practices needed to acquire and shape the workforce to the desired distribution.

- Procedure: Using an inventory projection model, the organization estimates the expected future inventory for each critical workforce characteristic (assuming continuation of current policies and practices) and presents the gap between that expected future inventory and the future desired distribution. Facilitated dialogue among the line leaders and human resource management participants identifies policies and practices necessary to eliminate the gaps (tying them to the critical characteristics and their desired distribution). The participants prioritize the policies and practices; discuss the need for revised data collection to support future workforce planning efforts; and outline a plan for securing necessary authority and resources and for implement-ing the policies and practices.

- Result: Agreement on a complete set of policies and practices needed to acquire and shape the workforce to the future desired distribution and the outline of how the organization will secure authority and resources and implement the policies and pro-grams.

REFERENCES

Burman, Allan V., Nathaniel M. Cavallini, and Kisha N. Harris, *Identification of the Department of Defense Key Acquisition and Technology Workforce, Washington*, DC: Jefferson Solutions, May 1999.

Corporate Leadership Council, *Workforce Planning*, Fact Brief, Washington, D.C., June 1997.

Corporate Leadership Council, *Workforce Planning*, Fact Brief, Washington, D.C., November 1998.

Ellinor, Linda, and Glenna Gerard, *Dialogue: Rediscover the Transforming Power of Conversation*, New York: John Wiley & Sons, Inc., 1998.

Hynes, Michael V., Harry Thie, John E. Peters, et. al., *Transitioning NAVSEA to the Future: Strategy, Business, Organization*, Santa Monica, Calif.: RAND Corporation, MR-1303-NAVY, 2002.

General Accounting Office, *Federal Employee Retirements: Expected Increase Over the Next 5 Years Illustrates Need for Workforce Planning*, Washington, D.C.: GAO-01-509, April 2001.

General Accounting Office, *Managing Human Capital in the 21st Century*, Washington, D.C.: GAO/T-GGD-00-77, March 2000.

Hamel, Gary, *Leading the Revolution*, Boston: Harvard Business School Press, 2000.

Hax, Arnoldo C., and Nicolas S. Majluf, *The Strategy Concept and Process: A Pragmatic Approach*, Upper Saddle River, N.J.: Prentice Hall, 1996.

Lindblom, Charles E., "The Science of Muddling Through," *Public Administration Review*, Vol. 19, 1959, pp. 79–88.

Lindblom, Charles E., "Still Muddling, Not Yet Through," *Public Administration Review*, Vol. 39, 1979, pp. 222–233.

Mintzberg, Henry, Bruce Ahlstrand, and Joseph Lampel, *Strategy Safari: A Guided Tour Through the Wilds of Strategic Management*, New York: Free Press, 1998.

National Academy of Public Administration, Center for Human Resources Management, *Building Successful Organizations: A Guide to Strategic Workforce Planning*, Washington, D.C., May 2000.

Office of Management and Budget, OMB Bulletin No. 01-07, May 8, 2001, Subject: Workforce Planning & Restructuring.

Office of the Secretary of Defense, *Shaping the Civilian Acquisition Workforce of the Future*, 2000, http://www.acq.osd.mil/dpap/ Docs/report1000.pdf, accessed June 23, 2003.

Ripley, David E., *Workforce Planning*, White Paper, Alexandria, Va: Society for Human Resource Management, June 2000.

Senge, Peter M., *The Fifth Discipline: The Art & Practice of he Learning Organization*, New York: Doubleday/Currency, 1990.